C000135164

450 numbered pages

GIANT NOTEBOOK

ruled paper | 8.5 x 11 in

this notebook belongs to :

by new style publisher
ISBN: 9798689242590